Agroecology's Promise

Sustainable Solutions for Food Production, Conservation, and Community Development

Eleanor Conrad

Table of Contents

1 Introduction to Agroecology

Agroecology isn't just another farming method; it's a fundamental shift in how we view food production. It's about understanding the intricate web of life within an agricultural system, recognizing that healthy soil, thriving biodiversity, and resilient ecosystems are crucial for sustainable food production. Instead of treating the farm as an isolated entity, agroecology integrates it into the broader landscape, acknowledging the interconnectedness of all living things. This holistic approach moves beyond simply producing food to encompass considerations of environmental health, social justice, and economic viability. It's about creating agricultural systems that not only feed people but also contribute to the overall well-being of the planet and its communities.

Unlike conventional agriculture's reliance on external inputs like synthetic fertilizers and pesticides, agroecology prioritizes working with nature. This means leveraging natural processes to enhance soil fertility, pest control, and nutrient cycling. Imagine a farm where diverse crops grow together, mimicking the complexity of a natural ecosystem. Beneficial insects thrive alongside the cultivated plants, naturally suppressing pest populations. The soil, rich in organic matter, teems with microbial life that contributes to nutrient availability and water retention. This isn't a romantic ideal; these are achievable goals backed by decades of research and practical application.

The principles of agroecology offer a clear framework for achieving these goals. These principles emphasize synergy, diversity, and resilience. Synergy highlights the interconnectedness of different components within the system—how improved soil health can lead to healthier plants, which in turn support beneficial insects, creating a positive feedback loop. Diversity, both in terms of crops and livestock, enhances stability and reduces vulnerability to pests and diseases. Resilience is the ability of the system to withstand and recover from shocks, such as drought or extreme weather events, something increasingly critical in a changing climate. These core principles directly translate into sustainable practices.

Implementing agroecological practices requires a shift in mindset, a move away from the industrial model of agriculture toward a more localized, people-centered approach. This doesn't imply a rejection of technology; it simply means selecting and applying technologies that support ecological balance and human well-being. Consider precision agriculture techniques, like using sensors to optimize irrigation and nutrient application, or using drones to monitor crop health. These technologies can be integrated into agroecological systems to enhance efficiency and reduce environmental impact, proving that technology and sustainability are not mutually exclusive.

Furthermore, agroecology recognizes the crucial role of social and economic factors in sustainable food systems. It emphasizes the importance of fair prices for farmers, equitable access to land and resources, and the empowerment of local communities. Farmers' markets, community-supported agriculture (CSAs), and local food networks are vital components of this approach, fostering direct relationships between producers and consumers. These connections strengthen food security, support local economies, and promote community development. This holistic approach underscores the interconnectedness of environmental, social, and economic sustainability, offering a far more complete solution than traditional methods.

Agroecology isn't just a theoretical framework; it's a practical approach being implemented successfully around the globe. From small-scale farms to larger-scale operations, agroecological principles are proving their effectiveness in enhancing food security, improving environmental quality, and creating resilient rural communities. By understanding the core principles of agroecology and applying these principles through diverse methods suited to individual contexts, we can actively contribute to creating a more sustainable and equitable food system for all. The future of food security rests not only on technological advancements, but also on a fundamental shift in our approach to agriculture—one that embraces the ecological complexity of food production and places human and environmental well-being at its center. The journey towards sustainable food systems starts with understanding and adopting the principles of agroecology. It demands a shift in mindset, a commitment to working with nature, and a recognition of the interconnectedness of all things.

1.1 Defining Agroecology

Agroecology isn't simply a farming method; it's a holistic approach that views agricultural systems as intricate ecosystems. Unlike conventional agriculture which often focuses on maximizing yield through external inputs, agroecology prioritizes the synergistic interactions within the farm environment itself, fostering resilience and sustainability. It's about understanding and working with nature, rather than against it, recognizing the interconnectedness of soil, plants, animals, and the human communities that depend upon them. This understanding is vital for creating truly sustainable food systems.

This integrated approach necessitates a shift in perspective. We must move beyond the narrow focus on singular crop production and embrace the complexity of biodiversity. Healthy agroecological systems thrive on a rich tapestry of species—crops, livestock, beneficial insects, and microorganisms—each playing a crucial role in maintaining balance and productivity. This diversity not only enhances resilience against pests and diseases but also increases nutrient cycling and soil health, making the system less reliant on external inputs such as fertilizers and pesticides.

Consider the difference between a monoculture field of corn and a diverse agroecological farm. The monoculture, while potentially yielding high short-term returns, is susceptible to widespread disease outbreaks and depletes soil nutrients. The agroecological farm, with its variety of crops and livestock, is far more resistant to such threats. Its diverse plant community attracts beneficial insects that help control pests naturally, while the livestock provide manure that enriches the soil, fostering a self-regulating, regenerative system.

The social dimension is equally crucial. Agroecology champions equity and justice within food systems, recognizing that agricultural practices impact not just the environment, but also the livelihoods and well-being of farmers and communities. It emphasizes community participation and knowledge sharing, fostering collaborations between producers, consumers, and researchers. Empowering local food systems through initiatives like farmers' markets and community gardens strengthens local economies and enhances food sovereignty. This is not just about producing food; it's about building resilient and equitable communities.

Moreover, agroecology directly addresses the challenges posed by climate change. By promoting carbon sequestration in soils, improving water use efficiency, and enhancing biodiversity, agroecological practices can help mitigate greenhouse gas emissions and adapt to the changing climate. The creation of diverse and resilient agroecosystems is essential for long-term food security in a world increasingly affected by climate change. This resilience is more than simply adapting to environmental stressors, it is about creating systems that actively absorb and mitigate those stresses.

Furthermore, the successful implementation of agroecological principles requires a shift in governance and policy. Supporting the transition towards agroecology demands changes in agricultural policies, incentives, and regulations. Governments play a pivotal role in providing financial and technical support to farmers transitioning to agroecological practices, promoting research and development, and fostering enabling environments for agroecological innovation. Effective policies can significantly enhance the adoption of agroecological practices and contribute to the building of a more sustainable food system.

Finally, effective agroecological practices often involve innovative approaches that combine traditional knowledge with modern scientific understanding. This means incorporating cutting-edge research into soil science, plant breeding, and pest management strategies, while also respecting and valuing the time-tested wisdom of generations of farmers. This collaborative approach fosters innovation and ensures that agroecological practices are both effective and appropriate to local contexts, allowing for sustainable agricultural practices to thrive across various communities and environments. The future of food security is inextricably linked to the success of agroecology.

1.2 Principles of Agroecology

Agroecology isn't merely a collection of farming techniques; it's a holistic philosophy guiding sustainable food production. At its core lies a deep understanding of ecological processes, recognizing the farm as a living ecosystem rather than a mere production unit. This interconnectedness demands a systems thinking approach, where every element, from soil microbes to pollinators, plays a vital role. Ignoring this interconnectedness risks ecological instability and compromises long-term productivity. We must move beyond simplistic, reductionist views to truly embrace the complexity of nature.

The principle of synergy is fundamental. Agroecological systems aim for a harmonious interplay between different components, creating a whole greater than the sum of its parts. For instance, integrating livestock with crop production—as in silvopastoral systems—provides mutual benefits. Livestock manure fertilizes crops, while crop residues feed the animals. This cyclical exchange reduces reliance on external inputs, bolstering resilience and minimizing environmental impact. Such integrated systems often exhibit superior productivity compared to monocultures.

Diversity underpins the resilience of agroecological systems. Monocultures, with their inherent uniformity, are vulnerable to pests and diseases. In contrast, a diverse array of crops, livestock breeds, and beneficial organisms creates a complex, robust system that can withstand environmental shocks. This biological diversity also enhances ecosystem services, including pollination, nutrient cycling, and pest control. The resulting increased stability and reduced reliance on external inputs make for a more sustainable and resilient agricultural model.

Effective agroecological practices require a detailed understanding of local contexts. What works in one region might not be suitable in another. Climate, soil type, and socio-economic factors all influence the optimal design and management of agroecological systems. This localized approach fosters innovation and adaptation, leading to context-specific solutions that are both environmentally sound and economically viable. We must therefore move away from standardized solutions and embrace the diversity of agricultural landscapes.

Equitable access to resources and knowledge is crucial. Agroecology must address social and economic inequalities that often hinder its adoption. Empowering farmers, especially women and marginalized communities, through access to appropriate training, technology, and markets is essential for achieving equitable outcomes. Furthermore, fostering collaboration and knowledge sharing amongst farmers and researchers is paramount for accelerating the transition to more sustainable practices. Truly sustainable food systems are only achievable when everyone has the opportunity to participate.

Continuous learning and adaptation are at the heart of agroecological practices. Regular observation and monitoring of the system are critical for identifying issues and adapting management strategies accordingly. This iterative process of learning from experience allows farmers to refine their practices and continually improve the ecological and economic performance of their farms. Agroecology is not a static set of rules but a dynamic process of ongoing refinement and innovation.

Agroecology is about more than just food production; it's about enhancing the well-being of people and the environment. It promotes biodiversity, enhances soil health, conserves water, and mitigates climate change. By working with nature, instead of against it, agroecology offers a path to sustainable food systems that are both productive and ecologically responsible. The future of food security and environmental stewardship depends on our embracing this holistic approach. A truly successful transition requires a paradigm shift, from a purely production-oriented approach to one that considers the long-term health of our planet and its inhabitants.

2 Sustainable Farming Practices

Crop diversification is the cornerstone of sustainable farming. Imagine a field bursting with not just one type of corn, but a vibrant mix of corn, beans, and squash – a classic example of the "three sisters" intercropping technique used by indigenous farmers for centuries. This method isn't just aesthetically pleasing; it actively builds soil health, boosts biodiversity, and reduces pest pressure, leading to a more resilient and productive system compared to monoculture farming. The diverse root systems of these crops capture nutrients from different soil depths, enriching the earth and reducing the need for synthetic fertilizers. Beans, with their nitrogen-fixing abilities, replenish the soil naturally, improving overall fertility. The squash acts as a ground cover, further suppressing weeds and improving water retention. The result? A healthier ecosystem and a higher yield overall, demonstrating a powerful strategy for long-term agricultural success and providing an ecological balance that traditional farming methods often lack. Integrated Pest Management (IPM) is another crucial component. Forget about spraying broad-spectrum pesticides that indiscriminately kill beneficial insects alongside the targeted pests. IPM focuses on a holistic approach. It starts with careful monitoring of pest populations; you learn to identify the specific pests affecting your crops and their life cycles. Then, you employ a combination of strategies: natural predators, resistant crop varieties, and targeted pesticide applications only when absolutely necessary. This approach minimizes environmental damage while maximizing crop protection. We can consider using pheromone traps to disrupt mating patterns, crop rotation to break pest life cycles, or encouraging the presence of beneficial insects like ladybugs that prey on aphids. This intricate system reflects a true partnership between human intervention and natural processes, emphasizing prevention over reaction. By carefully analyzing the ecosystem, you develop strategies far more effective and less harmful than blanket pesticide use.

Soil health is the bedrock of sustainable agriculture. Healthy soil is teeming with life: bacteria, fungi, earthworms, all working together in a complex web of interactions. These organisms improve soil structure, water retention, and nutrient cycling – naturally. No-till farming, cover cropping, and the addition of organic matter such as compost, all contribute to this vibrant soil ecosystem. By avoiding tillage, you preserve the soil structure, reducing erosion and enhancing water infiltration. Cover crops act as a living mulch, suppressing weeds, improving soil fertility, and preventing nutrient runoff. Compost adds organic matter, improving soil structure and nutrient availability. The long-term benefits are significant, including enhanced water-holding capacity, increased resistance to drought and disease, and reduced dependence on synthetic fertilizers. This, in turn, leads to greater environmental sustainability and a higher-quality product. Improving soil health requires an understanding of the diverse life within and should be considered a paramount step in any sustainable farming venture.

The importance of soil cannot be overstated; it is the foundation upon which all plant life depends. Treating soil health as a primary concern is an absolute requirement for the long-term health and success of any farm. Consider the role of biodiversity in a sustainable farm. It's not just about multiple crops; it's about creating habitat for a wide array of beneficial insects, birds, and other wildlife. Hedgerows, windbreaks, and other habitat features provide shelter and food sources for these creatures, enhancing natural pest control and pollination. A diverse farm ecosystem is more resilient to environmental stresses, diseases, and pest outbreaks, demonstrating a superior resilience compared to conventional farming methods. Think of it as a complex web: the more diverse and interconnected the web, the more resilient it is to external shocks. Supporting biodiversity is not merely an environmental imperative; it's a strategic choice that enhances farm productivity and sustainability, leading to a more robust and profitable enterprise. This interwoven network of life sustains the health of the entire system.

Water conservation is paramount. Sustainable farmers employ practices that minimize water use and maximize its efficiency. Efficient irrigation techniques, like drip irrigation, deliver water directly to the roots, reducing evaporation and runoff. Mulching helps retain soil moisture, reducing the need for frequent watering. Careful crop selection, choosing drought-tolerant varieties, is also crucial. We can also implement rainwater harvesting systems to capture and store rainwater for later use. It requires planning and resourcefulness but the rewards, in terms of both cost savings and environmental protection, are considerable. The judicious and responsible use of water in agriculture is not just good practice; it's a necessity for the long-term survival of both the farm and the wider ecosystem. Water is a precious resource; using it wisely is an ethical and economic imperative.

2.1 Crop Diversification

Planting a diverse array of crops is far more than simply scattering seeds; it's about strategically designing a resilient ecosystem on your land. Think of it like a well-balanced portfolio—diversification minimizes risk. If one crop fails due to disease, pests, or unfavorable weather, others will likely thrive, ensuring a stable harvest. This contrasts sharply with monoculture farming, where a single crop dominates, creating vulnerabilities across the board. Instead of relying on a single, potentially unstable income stream, a diverse approach provides multiple sources of revenue, and potentially extends your harvest season.

Consider the benefits that extend beyond mere yield. Planting a variety of crops with different root systems improves soil health. Some plants, like legumes, fix nitrogen in the soil, enriching it for other plants. Different root structures prevent soil erosion and improve water infiltration, boosting overall soil fertility and health. This approach helps to minimize the need for synthetic fertilizers.

Intercropping, a practice involving growing two or more crops in the same field at the same time, offers significant advantages. For example, combining a tall crop with a low-growing one can optimize light use and resource allocation, while companion planting can deter pests or attract beneficial insects, reducing the need for chemical pest control. This approach also lessens the need for herbicides, another significant benefit in the drive for more environmentally friendly approaches.

Beyond the purely practical aspects, crop diversification offers a chance to experiment and learn. It allows farmers to explore different market opportunities and tailor their production to local demands, whether it's supplying local restaurants or engaging in community-supported agriculture (CSA) programs. This direct connection with consumers creates a more equitable and sustainable food system. It empowers local farmers and supports local economies in ways a monoculture system simply cannot.

Furthermore, diversification allows for the integration of cover crops. These plants, grown specifically to improve soil health, are not harvested for food but play a crucial role in preventing soil erosion, suppressing weeds, and improving soil structure. Cover crops can also be incorporated into the crop rotation, adding another layer of resilience to the farm. This makes the entire operation more efficient.

Crop diversification also fosters biodiversity on your farm. A variety of plants supports a greater diversity of beneficial insects, birds, and other wildlife. This natural biodiversity strengthens the resilience of your farm against pests and diseases while creating a more attractive and ecologically diverse landscape. This approach fosters a healthier ecosystem and, naturally, a healthier farm.

However, transitioning to a diversified system isn't a simple switch. It necessitates careful planning and an understanding of local conditions. Farmers need to research appropriate crop combinations, consider soil types, assess market demands, and adapt their strategies to weather patterns. This planning stage is critical and should not be rushed. It requires the right expertise.

Investing time and resources in learning about different crops and their cultivation needs is essential. This might involve seeking advice from experienced farmers, attending workshops, or participating in educational programs. Collaboration with other farmers, sharing knowledge and experiences, can significantly enhance your success rate. Remember, sharing information makes for a better and safer agricultural environment.

Finally, remember that crop diversification is a journey, not a destination. It is an ongoing process of learning, adapting, and refining your practices. Farmers should regularly assess their progress, track the performance of different crops, and adjust their strategies as needed. This means adapting to a more dynamic environment, which provides more flexibility in the long term. Through continuous monitoring and experimentation, you can fine-tune your farm's ecosystem for maximal sustainability and productivity. This makes crop diversification more than just a farming practice; it becomes a long-term commitment to resilient agriculture.

2.2 Integrated Pest Management

Integrated pest management (IPM) is a cornerstone of sustainable agriculture, offering a holistic approach to pest control that minimizes reliance on synthetic pesticides. Instead of eradication, IPM prioritizes managing pest populations below economically damaging levels. This involves a detailed understanding of the pest's life cycle, its interactions within the ecosystem, and the economic threshold at which pest damage warrants intervention. Effective IPM necessitates careful monitoring, implementing preventative measures, and utilizing targeted interventions only when necessary.

Unlike conventional pest control methods that rely heavily on broad-spectrum pesticides, IPM emphasizes a multi-pronged strategy. This involves deploying a combination of techniques to control pests, taking into account the specific context of the farm. It's not a one-size-fits-all solution; the most effective approach will vary depending on factors like the crop being grown, the specific pests present, and environmental conditions. Careful consideration of these factors is crucial for success.

Cultural controls form a vital part of IPM. These methods focus on modifying the environment to make it less hospitable to pests. This includes simple practices such as proper crop rotation, selecting pest-resistant varieties, maintaining healthy soil, and ensuring appropriate irrigation and fertilization. These seemingly basic measures can drastically reduce pest populations and, importantly, improve the overall health and resilience of the agricultural system. They often work synergistically with other control methods.

Biological control utilizes natural enemies of pests to suppress their populations. This can involve introducing beneficial insects, such as ladybugs or lacewings, which prey on aphids or other crop pests. It also incorporates the use of nematodes, microscopic organisms that can effectively control soil-borne pests. Furthermore, strategically placed birdhouses or other wildlife habitat can attract natural predators, creating a more balanced ecosystem within the farm. The success of biological control hinges on understanding the intricate relationships between organisms within the agricultural environment.

Mechanical controls offer a straightforward approach to pest management. These include methods such as handpicking pests, using traps to capture specific insects, or deploying barriers to prevent pest access to crops. Physical removal of weeds that serve as pest hosts can also be highly effective. Simple, accessible methods such as these contribute significantly to an effective IPM strategy; their value shouldn't be underestimated. These low-impact solutions can be particularly useful in smaller scale operations.

Chemical controls, while part of IPM, are used judiciously as a last resort. If other methods fail to keep pest populations below the economic threshold, highly targeted insecticides may be employed. This necessitates thorough understanding of the specific pest, the proper application techniques, and potential environmental impacts. The careful, informed use of chemical controls differentiates IPM from conventional pesticide-heavy approaches, highlighting IPM's commitment to minimizing environmental harm. Careful record-keeping of all IPM strategies implemented is essential for evaluating their effectiveness and making informed decisions about future interventions.

Monitoring pest populations is a crucial component of IPM. Regular inspections of crops allow for early detection of pest infestations, before they reach damaging levels. This information is critical in determining the need for interventions, and what type of intervention is most appropriate. Methods for monitoring can range from simple visual inspections to the use of pheromone traps or other sophisticated monitoring tools depending on the pest and the resources available. Consistent monitoring provides valuable data for evaluating the overall effectiveness of the IPM strategy.

The successful implementation of IPM requires a comprehensive understanding of the ecological dynamics of the agricultural system. This entails studying the complex interactions between crops, pests, natural enemies, and environmental factors. A deep understanding of these interactions allows farmers to make informed decisions about pest management, selecting the most appropriate methods for each specific situation. It is an iterative process, constantly adjusting strategies based on observation and experience. Education and training are key to developing the necessary expertise for effective IPM. This means ongoing learning and information sharing

within the farming community.
Economic thresholds play a critical role in determining the necessity of intervention. These thresholds represent the pest population level at which the economic costs of pest damage outweigh the costs of control measures. The identification of economic thresholds is vital for balancing the costs and benefits of different control options and ensuring that interventions are both effective and economically sound. Understanding and applying this principle can prevent unnecessary or ineffective pesticide use. Accurate estimation of economic thresholds necessitates careful consideration of both the value of the crop and the costs associated with various control methods. This requires careful cost-benefit analysis.
Finally, successful IPM requires a collaborative approach. Farmers, agricultural extension agents, researchers, and other stakeholders must work together to share information, develop effective strategies, and adapt to changing circumstances. Effective IPM is a continuous learning process, requiring constant adaptation and refinement based on new data, improved understanding, and evolving needs. Strong community support and networking are vital to ensuring the long-term success and sustainability of IPM practices. The collaborative and adaptive nature of IPM makes it a cornerstone of sustainable agriculture for the future.

2.3 Soil Health Management

Healthy soil is the cornerstone of productive and sustainable agriculture. Ignoring soil health is akin to building a house on a crumbling foundation; the entire structure is at risk. Therefore, understanding and implementing effective soil health management practices is paramount for any aspiring agroecologist. This chapter will equip you with the knowledge and practical skills to nurture vibrant, fertile soil.

Soil is more than just dirt; it's a complex living ecosystem teeming with microorganisms, insects, and plant roots, all interacting in a delicate balance. This intricate web of life is vital for nutrient cycling, water retention, and overall soil fertility. Disrupting this balance, through intensive tillage or excessive chemical inputs, can lead to soil degradation and reduced crop yields. The health of the soil directly impacts the health of the plants that grow from it and ultimately the health of your food supply. Consider the simple action of stepping onto the soil and compare it to walking on pavement; your step does not affect the pavement, but walking on soil can potentially impact the soil structure and soil life, which is why minimizing soil disturbance is essential. This can be achieved through the implementation of conservation tillage techniques such as no-till farming.

Building soil organic matter is crucial for enhancing soil health. Organic matter, composed of decomposed plant and animal residues, acts like a sponge, improving water retention and reducing erosion. It also provides a food source for beneficial soil organisms, further enhancing nutrient cycling. Cover cropping, a technique involving planting specific crops to protect the soil between cash crops, is an effective method for increasing organic matter content. Consider incorporating a mix of legumes, grasses, and other cover crops to maximize the benefits. Furthermore, the addition of compost and manure significantly increases organic matter levels and boosts soil fertility. Think of the soil like a living organism; you need to feed it and care for it to thrive.

Effective soil health management necessitates a move away from intensive tillage practices. Conventional tillage methods, which involve repeatedly plowing and turning the soil, disrupt the soil structure, damage soil organisms, and contribute to soil erosion. No-till farming, a conservation tillage method, helps to maintain soil structure, reduce erosion, and improve water infiltration. By leaving crop residues on the soil surface, no-till farming protects the soil from the harsh effects of sun and wind. This approach also creates a more favorable environment for beneficial soil organisms. Implementing no-till is not always simple; it might require adjustments in your planting and harvesting methods. It can be a gradual transition, starting with some areas of your land and then expanding it as you improve your technique.

Protecting soil from erosion is another critical aspect of soil health management. Erosion, the process of soil particles being carried away by wind or water, leads to significant loss of topsoil, which contains the most valuable nutrients. Implementing conservation tillage practices, such as no-till farming and contour farming, significantly reduces erosion. Proper water management, including the construction of terraces and swales, further mitigates erosion risks. Consider planting windbreaks and cover crops to help stabilize the soil. Additionally, buffer strips along the edges of fields can help reduce runoff. Planning your planting strategy considering potential erosion is a fundamental step in ensuring long-term soil health.

The use of cover crops can have multiple benefits. The roots of these plants, left in the soil after they have matured, improve soil structure and water infiltration. They also suppress weeds, reduce nutrient leaching, and contribute to soil organic matter. Furthermore, cover crops can improve the diversity of plant life in your fields, which can have broader positive impacts on the entire ecosystem of your farm. There are multiple cover crop options to choose from, and picking the right ones for your soil conditions and climate is essential for maximizing the beneficial effects. You should conduct a thorough analysis of your soil type, local climate, and your crop rotation planning to select the most appropriate ones.

Finally, regularly monitoring your soil health is crucial for effective management. Soil testing is an important tool for assessing nutrient levels, pH, and other vital parameters. Based on the test results, you can adjust your fertilizer and amendment applications to ensure optimal soil fertility. By paying close attention to the condition of your soil and making informed adjustments to your management practices, you can cultivate healthy, productive soil for years to come. Remember that soil health is a continuous process; it requires consistent attention and thoughtful decision-making. This holistic approach is essential for the long-term success of any agroecological system. Regular observation and meticulous record-keeping contribute significantly to achieving the desired outcomes, providing valuable insights for improving your methods over time.

3 Conservation Agroecology

Conservation agroecology goes beyond simply producing food; it actively safeguards our natural resources. It's a holistic approach, weaving together biodiversity protection with efficient water management to create truly sustainable agricultural systems. Think of it as designing a farm that not only feeds people but thrives as part of a healthy ecosystem.

This isn't about isolated actions; it's about creating synergistic relationships within the farm environment. Planting diverse crops creates habitats for beneficial insects, reducing the need for pesticides and simultaneously increasing pollination. This diversification not only boosts yields but also safeguards against crop failures due to disease or pest outbreaks, crucial elements in resilient farming. The interwoven nature of the system ensures robust productivity even in the face of climatic variability.

Implementing conservation agroecology means proactively managing water resources. Techniques like rainwater harvesting and drip irrigation reduce water waste, conserving precious resources, especially vital in arid or semi-arid regions. The goal isn't just to irrigate efficiently; it's to improve water infiltration into the soil. This reduces runoff, protects water quality, and replenishes groundwater reserves; all of which contribute to long-term environmental sustainability.

We must consider the crucial role of hedgerows and windbreaks in conservation agroecology. These aren't mere aesthetic additions; they provide crucial habitat for wildlife, acting as corridors for movement and breeding grounds for beneficial insects and birds. Furthermore, they play a vital role in soil conservation by reducing wind erosion and regulating microclimates within the field. Their presence enriches the biodiversity of the farm, fostering a vibrant and resilient ecosystem.

Soil health is the bedrock of conservation agroecology. Avoid intensive tillage; instead, practice no-till farming or minimum tillage. This preserves soil structure, reducing erosion and promoting healthy microbial communities. These microorganisms are essential for nutrient cycling and improve soil fertility, minimizing the need for synthetic fertilizers. Healthy soils are more resilient to droughts, floods, and other climate-related stresses; enhancing overall farm productivity and environmental stability.

Integrated pest management (IPM) is an essential component of a conservation agroecology strategy. IPM emphasizes prevention and natural control methods over the use of synthetic pesticides. This includes crop rotation, companion planting, biological control agents (like beneficial insects), and monitoring pest populations. This approach protects beneficial insects and pollinators, maintains the integrity of the ecosystem, and reduces the environmental impact of farming. It's about building resilience, making the farm less vulnerable to pest infestations.

Conservation agroecology isn't merely a collection of techniques; it requires a shift in mindset. Farmers must become keen observers of their farms, understanding the interactions between crops, soil, water, and the surrounding environment. By learning to read these signals, farmers can make informed decisions that promote ecosystem health and maximize the farm's resilience and productivity. This observational approach is integral to creating a genuinely sustainable system.

This understanding extends beyond the farm itself. Conservation agroecology necessitates collaboration. Farmers need to engage with researchers, policymakers, and their communities to share knowledge, develop strategies, and create policies that support sustainable agricultural practices. The transition to conservation agroecology requires a collective effort; it's a community project, not simply an individual one. Success depends on shared goals and mutual support within agricultural networks.

Integrating conservation agroecology effectively also requires addressing economic concerns. While initially it might demand more labor or initial investment, the long-term benefits—reduced inputs, increased yields, improved resilience—often outweigh the costs. Furthermore, the positive environmental impact can lead to new market opportunities and enhance the value of sustainably produced crops. The economic benefits should be considered alongside the environmental ones, ensuring the approach is both environmentally and economically viable.

By embracing the principles and practices of conservation agroecology, farmers can build thriving, resilient, and ecologically sound farms that provide food while contributing to a healthier planet. It's about creating a synergy between production and conservation, fostering a system that benefits both people and nature. The future of food production depends on integrating these principles, creating a sustainable agricultural system for generations to come. This isn't just about farming; it's about stewardship of our shared environment.

3.1 Biodiversity Conservation

Biodiversity underpins the resilience of agroecosystems. A diverse range of plant species, for instance, contributes to greater overall productivity and stability by providing a wider array of nutrients, resisting pests and diseases more effectively, and improving soil structure. The presence of numerous plant species allows for the creation of complex food webs, enhancing the ecosystem's natural pest control mechanisms and reducing reliance on external inputs like pesticides. This diversity also provides a crucial buffer against the unpredictable impacts of climate change. Different species exhibit varied tolerances to drought, heat, or disease outbreaks; a diversified system is thus less susceptible to catastrophic failures.

Healthy soils are fundamental to biodiversity. Maintaining robust soil biodiversity, which includes diverse populations of microbes, fungi, and invertebrates, is essential. These organisms contribute significantly to nutrient cycling, decomposition, and soil aggregation, processes which are crucial for plant growth and overall ecosystem health. This, in turn, supports a wider array of plant life and associated fauna, fostering greater biodiversity. Practical steps to improve soil health include reducing tillage, employing cover cropping, and integrating livestock grazing to enhance organic matter content, water retention, and microbial activity. Avoid monocultures and promote intercropping strategies which reduce susceptibility to pests and diseases, while supporting a richer soil community.

Protecting pollinators is paramount. Pollinators, including bees, butterflies, and other insects, are essential for the reproduction of many plant species, both wild and cultivated. Their decline poses a significant threat to agricultural production and biodiversity. Agroecological practices play a vital role in pollinator conservation by providing diverse habitats with nectar-rich flowering plants and nesting sites. Reducing pesticide use is crucial, as many pesticides are highly toxic to pollinators. Creating hedgerows and other habitat features within and around agricultural fields can significantly improve pollinator populations and enhance the overall biodiversity of the landscape. Furthermore, diversifying crops and promoting natural pest control strategies can further reduce pesticide use. A carefully planned, holistic approach is necessary for successful pollinator protection.

Wildlife habitats within agroecosystems are crucial for biodiversity maintenance. The integration of natural habitats such as hedgerows, riparian buffers, and woodland strips provides refuge for a wide range of wildlife species, including birds, mammals, reptiles, and amphibians. These habitats play crucial roles in supporting biodiversity by offering nesting sites, foraging areas, and corridors for movement, connecting different habitats within the landscape. This connectivity is essential for maintaining viable populations of many species and supporting gene flow between populations. Designing agricultural landscapes to incorporate these natural habitats is vital for enhancing biodiversity and contributes significantly to ecosystem services, such as pollination, pest control, and nutrient cycling. This creates a more resilient and sustainable agricultural system.

Water management practices can enhance biodiversity. Efficient irrigation techniques, such as drip irrigation, can help reduce water consumption, while conserving water resources for both agricultural production and the wider ecosystem. Water conservation measures are important to protect aquatic habitats, supporting a rich diversity of plant and animal life dependent on these resources. The preservation of wetlands and other aquatic habitats is also important for maintaining biodiversity, as these areas often support a high number of species. Protecting water quality by minimizing the use of fertilizers and pesticides helps to maintain aquatic biodiversity and prevent negative impacts on downstream ecosystems. Sustainable water management is a crucial component of successful agroecological systems. These multiple components contribute to an interconnected and sustainable farming model.

3.2 Water Resource Management

Efficient water management is paramount in agroecological systems. Unlike conventional agriculture, which often relies on intensive irrigation and depletes groundwater reserves, agroecology emphasizes water conservation through methods that mimic natural water cycles. This involves careful consideration of soil type, topography, and local rainfall patterns to optimize water use efficiency. Understanding these factors allows for the implementation of appropriate water harvesting techniques and irrigation strategies, minimizing water waste and maximizing crop yields.

The choice of crops plays a crucial role in water management. Drought-tolerant varieties, for example, require less irrigation, reducing reliance on external water sources. Similarly, selecting crops suited to the specific soil and climate conditions minimizes water stress and optimizes growth. Crop rotation, a key agroecological practice, improves soil water retention capacity by enhancing the soil's organic matter content. This in turn minimizes runoff and improves infiltration, making water available to plants for longer periods.

Intercropping, or growing multiple crops simultaneously, can also enhance water use efficiency. This is due to the synergistic effects of different plants on the soil and overall ecosystem. Certain crops can improve water infiltration and reduce evaporation, improving the water availability for other plants in the system. This synergistic effect reduces the need for supplemental irrigation and improves overall resource use efficiency within the system. It demonstrates the importance of understanding plant interactions in water resource management within agroecological systems.

Careful land management practices are essential for successful water conservation. Terracing, contour farming, and other techniques minimize soil erosion, thereby preventing the loss of precious topsoil and improving water infiltration. These practices can significantly reduce water runoff, allowing more water to be absorbed by the soil and used by plants. Implementing effective soil conservation strategies is essential in protecting the soil's ability to retain water and nutrients.

Beyond the farm level, effective water resource management requires collaboration and community engagement. This involves sharing knowledge and resources among farmers, promoting water-efficient irrigation technologies, and developing community-based water management plans. Working together to ensure equitable access to water resources and implement responsible water-use strategies is crucial for sustainability and community well-being. Community involvement ensures that water management strategies are appropriate and effective for local conditions.

The integration of rainwater harvesting systems is a crucial aspect of agroecological water management. This involves collecting and storing rainwater for later use during dry periods, reducing dependence on external water sources, and mitigating the impact of droughts. These systems, whether simple or complex, play a crucial role in ensuring a reliable water supply for irrigation and other farm needs, reducing the pressure on groundwater resources.

Monitoring soil moisture levels is critical for efficient irrigation scheduling. Using soil moisture sensors allows farmers to precisely determine when and how much water is needed, minimizing water waste and optimizing plant growth. This data-driven approach reduces water consumption while maintaining optimal crop yields, demonstrating the integration of modern technology for sustainable resource management. This technology improves the accuracy and efficiency of water allocation within agroecological systems.

Considering the impact of climate change is crucial in water resource management. Changes in rainfall patterns and increased frequency of droughts necessitate the adoption of resilient water management strategies. This includes diversifying water sources, investing in water-efficient irrigation systems, and developing drought-resistant crops to buffer against climate change's impacts. Preparing for these challenges ensures long-term water security within the agroecological system.

Agroecology provides a comprehensive approach to water resource management, integrating ecological principles with practical farming techniques. By carefully considering the interplay between soil, plants, and water, farmers can create resilient and productive systems that minimize water use while maximizing yields. This holistic approach ensures sustainable water management in the long term, benefiting both the environment and the farming community. Careful planning and thoughtful implementation are essential for the long-term success of any agroecological water management strategy.

4 Community Food Systems

Community food systems are more than just farmers' markets; they represent a fundamental shift in how we produce, distribute, and consume food. They prioritize local food production, fostering strong connections between farmers and consumers, and creating resilient and equitable food access within communities. This fosters a sense of place and strengthens community bonds in ways that mass-produced food systems simply cannot.

Building a thriving community food system necessitates a multi-pronged approach. It demands a strong network of local farmers, access to land suitable for agriculture, and innovative approaches to food distribution. Farmers markets, though visibly impactful, are only one component. Community-supported agriculture (CSA) programs, where consumers purchase shares of a farm's harvest in advance, provide a vital link, guaranteeing the farmer income while providing consumers with fresh, seasonal produce. Collaborations between farmers, restaurants, and institutions – such as schools and hospitals – can create robust distribution channels, ensuring that local food reaches a wider population. This interdependence offers economic stability and food security for all participants.

Food security, at its core, is about ensuring access to enough safe and nutritious food for everyone, regardless of income. Within the framework of community food systems, food security means not just supplying food but ensuring affordable access. This is where initiatives like food banks and community gardens come into play, providing a safety net for those struggling with food insecurity. Community gardens, in particular, act as vital educational tools, teaching valuable skills while fostering a deeper connection with food production and biodiversity. They are spaces where people of all backgrounds can come together, share knowledge, and cultivate a sense of collective responsibility for their community's well-being. Beyond supplying food, they are vital social hubs, fostering connection and understanding.

Furthermore, advocating for policies that support local food systems is crucial. This includes supporting initiatives that prioritize local food procurement for public institutions, such as schools and hospitals, and incentivizing farmers to grow food for their communities. It also involves policies that protect farmland from development, ensuring its long-term availability for food production. Strong policies that ensure fair prices for farmers and protect consumers from exploitative practices are essential components of creating a truly just food system. It isn't merely enough to have access to food; that access must be equitable and accessible to everyone.

Effective community food systems go beyond mere economic transactions. They are about fostering a deep sense of community ownership and shared responsibility. They create opportunities for education, skill-building, and engagement. Workshops on food preservation, cooking classes using local ingredients, and programs that connect schoolchildren with local farms can foster a deeper understanding of food systems and cultivate an appreciation for healthy, sustainable eating habits. This holistic approach promotes healthier lifestyles and strengthens community ties across generational and

economic divides. This reciprocal understanding and shared responsibility is, quite simply, essential for a vibrant and resilient community.
The creation of a robust community food system requires sustained effort and collaboration. It demands active participation from individuals, organizations, and local governments. By adopting a holistic approach that integrates agricultural production, food distribution, food justice, and community engagement, we can create food systems that are both ecologically sound and socially equitable. These systems will ultimately foster more resilient, sustainable, and thriving communities for generations to come.

4.1 Local Food Networks

Local food networks are crucial for building resilient and sustainable food systems. They foster direct connections between producers and consumers, strengthening community bonds and promoting economic development within the local area. Imagine a farmer, Sarah, who cultivates heirloom tomatoes using organic methods. Instead of relying on large-scale distributors, Sarah sells her produce directly to local restaurants, grocery stores, and consumers through a community-supported agriculture (CSA) program, ensuring fair prices and fresh, high-quality food for her community. This direct relationship builds trust and transparency, allowing consumers to understand the origins of their food and support sustainable practices.
This model reduces food miles, the distance food travels from farm to table, significantly lowering the environmental impact associated with transportation. Greenhouse gas emissions are reduced, and the reliance on fossil fuels is minimized. The freshness of the produce is also vastly improved, leading to less food waste and a higher nutritional value for consumers. Moreover,

local food networks often support smaller, diverse farms, thereby preserving biodiversity and traditional agricultural knowledge. They challenge the dominance of large, industrialized agriculture, which frequently utilizes unsustainable practices and contributes to environmental degradation.

The economic benefits of local food networks extend beyond the farmers themselves. Local businesses, such as restaurants and grocery stores, benefit from sourcing fresh, high-quality ingredients, enhancing their reputation and attracting customers who value sustainability and community engagement. The creation of jobs within the local food system strengthens the local economy and creates opportunities for community members. Processing and distribution roles often emerge within these networks, diversifying local employment opportunities beyond farm labor. Increased consumer spending remains within the community, leading to economic growth and increased financial stability for the local area.

Beyond the economic advantages, local food networks contribute to community building and social equity. They foster social connections, encouraging community interaction and creating spaces for sharing knowledge and skills. Community-supported agriculture (CSA) programs, farmer's markets, and local food initiatives often become central meeting places, enhancing social cohesion and community spirit. These networks can also provide access to healthy food for underserved populations, addressing issues of food insecurity and promoting food justice. By supporting local food producers, communities demonstrate a commitment to equitable food access, especially important for vulnerable populations who may lack access to fresh, affordable food.

Participating in local food networks requires an active commitment from both producers and consumers. Producers need to build strong relationships with buyers, efficiently manage their harvests, and meet the demands of the local market. This might require adapting their farming practices to suit the specific needs of their community and developing effective marketing strategies to reach their target customers. Consumers, in turn, need to actively seek out locally produced food, understand the challenges faced by local farmers, and be willing to pay fair prices that support sustainable agricultural practices. Choosing local food is an act of mindful consumption that extends beyond personal benefit to the greater well-being of the community and environment. The collaborative effort involved in creating thriving local food networks is essential to building a more sustainable and equitable food future.

Furthermore, supporting local food systems helps to preserve cultural heritage and traditions. Many local foods represent centuries-old farming practices and recipes unique to particular regions. These traditions and practices are often at risk due to the globalization of food systems, which favor mass-produced, standardized products. Local food networks provide a vital platform for preserving this cultural heritage and sharing traditional culinary knowledge. Farmers' markets, for instance, can act as vibrant showcases of local cuisine, providing valuable opportunities for education and cultural exchange. The direct connection between producers and consumers allows for the sharing of stories and knowledge, thus preserving valuable cultural heritage for future generations. This not only enriches community life but also adds layers of flavor and meaning to the food itself.

However, establishing and maintaining robust local food networks requires careful consideration of various challenges. Seasonal availability of produce may limit market diversity, requiring creative strategies for preserving and storing surplus food. The scaling up of local production may be limited by land availability or access to appropriate resources. Moreover, local food networks must address issues of equitable access and affordability, ensuring that all members of the community benefit from this system. Collaboration with local organizations, government support, and innovative strategies for addressing these challenges are critical to creating truly sustainable and inclusive local food networks. This requires collaborative efforts among farmers, consumers, policymakers, and community organizations to overcome these challenges and create thriving local food systems for the benefit of all.

4.2 Farmer's Markets & CSAs

Farmer's markets and Community Supported Agriculture (CSA) programs represent powerful tools for building strong, resilient local food systems, core components of any thriving agroecological approach. They foster direct connections between farmers and consumers, promoting transparency and community engagement in ways that large-scale industrial agriculture simply can't replicate. This direct connection helps build trust, strengthening the entire local food economy. It also allows for valuable feedback loops, where farmers can learn directly from consumers about their preferences, while consumers gain a deeper understanding of farming practices.

Participating in or establishing a farmers' market offers numerous benefits to both farmers and consumers. For farmers, it provides a valuable outlet for selling their produce directly to consumers, bypassing the often-expensive and exploitative middlemen of conventional supply chains. This direct access to market translates to higher profits and increased control over pricing strategies, empowering farmers economically and contributing to a more equitable agricultural landscape. Consumers, in turn, gain access to fresher, higher-quality produce, often grown using sustainable and environmentally friendly methods. The ability to interact directly with the farmers who grew their food fosters a greater appreciation for the labor and effort involved in food production, something often obscured in the anonymity of large supermarkets.

Community Supported Agriculture (CSA) operates on a slightly different, but equally valuable, model. CSA programs typically involve consumers purchasing a "share" of a farm's harvest in advance, receiving regular deliveries of fresh produce throughout the growing season. This subscription-based model provides farmers with a guaranteed income stream, reducing the risks inherent in relying solely on fluctuating market prices. It also encourages a longer-term commitment between farmers and consumers, solidifying their relationship and building a sense of shared responsibility for the local food system. For consumers, CSAs offer the opportunity to build relationships with their food sources and gain access to a wider variety of seasonal produce they might not find in a typical grocery store. The regular deliveries also introduce people to unfamiliar fruits and vegetables, increasing their dietary diversity and improving overall health.

However, both farmers' markets and CSAs face unique challenges that need careful consideration. Farmers' markets often require significant time and effort to set up, maintain, and manage, from securing permits and licenses to handling logistics and marketing. The seasonal nature of farming can also impact consistency of supply, requiring farmers to plan carefully and perhaps diversify their offerings to maintain customer interest throughout the year. For CSAs, challenges can include managing member communication, ensuring timely deliveries of high-quality produce, and maintaining fair pricing strategies that both compensate farmers adequately and remain affordable for consumers. Managing member expectations and handling issues such as unexpected crop failures are also integral to successfully operating a CSA. Successful operation often necessitates robust logistical planning and flexible approaches to address unforeseen circumstances. Successful implementation of both farmer's markets and CSAs requires careful planning, community engagement, and a focus on sustainable practices. Promoting the benefits of locally-sourced food, educating consumers about the benefits of sustainable agriculture, and ensuring access to these programs, particularly for low-income communities, are crucial for creating truly inclusive and resilient local food systems. Collaboration between farmers, consumers, and local governments is essential in overcoming these challenges and building a strong foundation for agroecological food production within a thriving, engaged community. Moreover, understanding the interplay between local governance, market economics, and consumer behavior is vital to maximizing the impact of farmer's markets and CSAs. These programs represent not merely economic transactions, but rather the foundation of a robust and connected food system, bolstering local economies and fostering community resilience. A well-functioning farmers market or CSA fosters a virtuous cycle, benefiting all parties involved.

4.3 Food Security & Justice

Food security, at its core, isn't just about having enough food; it's about ensuring everyone has access to nutritious, culturally appropriate food. This requires a fundamental shift away from industrial agriculture's emphasis on profit maximization towards a more equitable and sustainable system.

Agroecology provides the framework for this transition, but achieving true food security necessitates addressing the systemic injustices that currently impede access to healthy food for vulnerable populations. We must acknowledge the historical and ongoing impact of colonialism, land dispossession, and discriminatory policies on food systems globally. These injustices have disproportionately affected marginalized communities, leaving them with limited access to land, resources, and markets.

The fight for food justice is intertwined with broader struggles for social and environmental justice. Empowering marginalized farmers and communities is crucial. Providing them with the resources, training, and support they need to adopt agroecological practices is essential, rather than viewing them merely as recipients of aid. We need to foster self-determination and build their capacity to participate actively in shaping the food systems that sustain them. This means focusing on education, technical assistance, and equitable access to credit and markets. Investing in community-based organizations, farmer cooperatives, and other local initiatives is crucial in this effort.

Supporting local food systems is fundamental to achieving both food security and justice. Local food networks, farmers' markets, and community-supported agriculture (CSAs) build more resilient and equitable food systems by fostering direct connections between producers and consumers. This approach strengthens local economies, improves access to fresh produce, and supports environmentally sustainable farming practices. Local markets provide opportunities for farmers, particularly those from marginalized communities, to receive fair prices for their products while circumventing exploitative intermediaries. Furthermore, empowering these farmers to lead in local food initiatives promotes self-reliance and community development.

Community gardens are vital components of food justice initiatives. They not only provide access to fresh, healthy food, especially in food deserts, but also serve as crucial spaces for community building and education. These gardens offer opportunities for skill-building, promoting healthy eating habits, and fostering a deeper connection between people and their food. Furthermore, they often function as sites of social activism and community empowerment, challenging inequalities and advancing social justice agendas. They demonstrate the power of collective action in creating more equitable and resilient food systems.

Addressing food insecurity and injustice demands a multifaceted approach. We must challenge existing power structures and promote policies that prioritize food sovereignty and environmental sustainability over corporate profits. This includes advocating for policies that support local food systems, protect small farmers, promote agroecological practices, and ensure equitable access to land and resources. Building robust food policy and advocacy movements at both local and national levels is pivotal. Mobilizing community support and political will is necessary to change unjust food systems.

Fair trade practices are also vital to ensuring food justice. They guarantee that farmers in developing countries receive fair prices for their products, enabling them to improve their livelihoods, invest in sustainable farming practices, and secure their food security. This contrasts sharply with conventional trade models that frequently exploit farmers, leading to poverty and environmental degradation. Supporting fair trade organizations and advocating for policies that promote fair trade standards plays a crucial role in fostering economic justice within global food systems.

The path to food security and justice is a journey, not a destination. It requires constant vigilance, adaptation, and a commitment to collective action. It's essential to understand and address the root causes of food insecurity and injustice and build systems that truly serve the needs of all people and the planet. The focus should always remain on ensuring equitable access to healthy, culturally appropriate food for everyone. This requires a fundamental shift in our priorities, values, and approaches to food production, distribution, and consumption. Agroecology, with its focus on ecological principles and social equity, offers a crucial pathway towards a just and sustainable food future. But it's only a pathway; the effort must come from individuals, communities, and governments working together.

4.4 Community Gardens

Community gardens offer a powerful blend of ecological restoration and social empowerment. They're not just about growing food; they're about fostering a sense of belonging, enhancing community bonds, and providing access to fresh, healthy produce. Imagine a neighborhood where residents, regardless of their background or experience, come together to cultivate a shared space. This shared experience transcends the simple act of planting and harvesting; it builds bridges between people, fostering understanding and collaboration.

Successfully establishing a community garden requires careful planning and community involvement from the very beginning. It's crucial to identify a suitable location—one with adequate sunlight, access to water, and good soil quality. But equally important is securing the support and participation of the local community. Public meetings, surveys, and online forums can help gauge interest, identify potential volunteers, and address any concerns residents might have. A well-defined governance structure, with clear roles and responsibilities for different tasks, is also essential for the garden's long-term sustainability.

The initial phases involve soil preparation, which might include testing for nutrient levels and addressing any soil contamination issues. This is a crucial step in ensuring healthy plant growth and reducing the need for chemical fertilizers. Building raised beds or utilizing existing structures can provide a more controlled gardening environment, particularly important in areas with less-than-ideal soil conditions. Selecting appropriate plant varieties – those that are well-suited to the local climate and soil conditions – is equally critical. Choosing a diverse range of plants can ensure a consistent harvest throughout the growing season and attract beneficial insects, promoting natural pest control.

Successful community gardens necessitate a commitment to sustainable practices. Composting kitchen scraps and yard waste provides a rich source of organic matter, improving soil fertility while reducing waste. Water conservation techniques, such as drip irrigation or rainwater harvesting, are also important to minimize water usage and protect local water resources. Integrated pest management (IPM) strategies, prioritizing natural pest control methods over chemical pesticides, ensures the safety of both the produce and the surrounding environment. Furthermore, regular soil testing and analysis will help gardeners to adapt their practices as needed.

Beyond the practical aspects of gardening, community gardens offer invaluable educational opportunities. Workshops and training sessions on various gardening techniques can empower residents with the knowledge and skills needed to cultivate their own gardens. These workshops can cover topics ranging from basic planting and harvesting to advanced techniques like seed saving and composting. By promoting sustainable practices, these gardens transform into living laboratories where people can learn about ecological balance, biodiversity, and the interconnectedness of the natural world. Educational programs can also focus on nutrition and healthy eating habits, linking the garden's produce directly to improved health outcomes within the community.

The benefits extend beyond education; community gardens foster a strong sense of community ownership and pride. The collaborative nature of the project strengthens social connections and improves the overall wellbeing of participants. The shared responsibility for maintaining the garden cultivates a sense of collective achievement, boosting morale and fostering social cohesion within the community. The social interactions during working sessions, from planting seeds to harvesting crops, create opportunities for people of different ages and backgrounds to connect, share knowledge and build meaningful relationships. This kind of social interaction is often missing in modern urban environments.

Moreover, community gardens contribute significantly to food security, especially in food deserts. They provide a source of fresh, healthy produce for residents, improving their access to nutritious food and reducing their reliance on processed foods. This is particularly important in low-income communities, where access to affordable and healthy food options is often limited. Distributing the harvested produce through community sharing programs or farmers' markets also creates economic opportunities for residents and boosts the local economy. The garden itself becomes a hub of activity, attracting visitors and creating a more vibrant and engaging community space.

In conclusion, community gardens offer a multifaceted approach to sustainable food production. They are not simply places to grow food; they are living examples of sustainable agriculture practices, educational centers promoting ecological awareness, and vibrant community hubs fostering social cohesion and improving food security. By engaging in the planning, implementation and maintenance of a community garden, individuals learn practical skills, forge lasting connections and actively contribute to a healthier, more sustainable and socially connected community. The collaborative and educational nature of these initiatives makes them powerful tools for promoting community development, environmental stewardship, and food security, all core principles of agroecology.

4.5 Food Policy & Advocacy

Effective food policy and advocacy are crucial for transitioning to agroecological systems. It's not merely about growing food sustainably; it's about creating a supportive environment where such practices thrive. This requires a multi-pronged approach, engaging various stakeholders and influencing policy decisions at local, regional, and national levels.

Successful advocacy hinges on building coalitions. Farmers, consumers, scientists, policymakers, and community groups—all possess unique perspectives and strengths that, when combined, create a powerful force for change. Finding common ground, understanding shared goals, and fostering collaborative relationships are paramount. Remember, a unified voice carries far more weight than isolated efforts. Building these coalitions requires effective communication, trust-building, and a willingness to compromise. It's about shared vision, not individual agendas.

Direct engagement with policymakers is key. This means attending public hearings, lobbying elected officials, and submitting written testimony to relevant committees. It involves clearly articulating the benefits of agroecology, emphasizing its role in improving food security, environmental sustainability, and rural economic development. Present data demonstrating the economic and environmental viability of agroecological practices, backed by scientific research and real-world case studies. Highlight success stories of agroecological farms in your region, providing concrete examples of their positive impact. This makes your message relatable and impactful.

Public awareness campaigns are equally important. Educating consumers about the benefits of agroecological food, its environmental impact, and the importance of supporting local farmers directly influences market demand. Employ diverse communication channels—social media, workshops, farmers' markets, educational materials, and collaborations with local media—to reach a wide audience. Effective storytelling, showcasing the human side of agroecology, can make a profound impact. Share the stories of farmers who have successfully transitioned to agroecological practices, detailing their challenges and successes. Let their experiences speak to the realities of agroecological farming.

Supporting policies that promote agroecology demands consistent effort. This includes advocating for subsidies and tax incentives for agroecological farmers, promoting research and development in agroecological techniques, and working towards regulations that support sustainable food systems. Analyze existing agricultural policies, identifying gaps and proposing amendments to better align with agroecological principles. Remember to support the development of farmer-led agroecological research networks; participatory research methods empower farmers and produce more relevant results. This collaborative approach ensures that policies accurately reflect on-the-ground needs and experiences.

Local food policy councils can become vital allies in this effort. These councils bring together various stakeholders to address local food system challenges, creating opportunities for collaborative advocacy on issues such as food access, farmers' markets, and urban agriculture. These councils offer a structured avenue for grassroots initiatives, providing a platform to directly influence local policy and resource allocation. Their active engagement helps align local policies with national and regional goals and objectives.

International collaboration is becoming increasingly vital. Connecting with international organizations, attending international conferences, and participating in global food policy discussions broaden perspectives and identify best practices and successful strategies from other regions. Learning from global experiences and sharing successes strengthens the global agroecology movement, fostering a collaborative effort towards a more sustainable and equitable food system. This also allows for global sharing of research and methodologies, facilitating faster progress across regions.

Finally, remember that advocacy is a long-term commitment. It demands persistence, adaptability, and a willingness to learn and evolve. Celebrate small victories, recognizing the incremental progress that contributes to the overall success of the movement. Build resilience in the face of setbacks, understanding that significant change requires time and concerted effort. Remember that building a strong foundation of support will enable long-term success. Continuous refinement of your advocacy strategies and adaptation to the ever-changing political landscape will solidify its impact. The journey toward a more sustainable and just food system is a marathon, not a sprint.

5 Agroecology and Climate Change

Climate change poses a significant threat to global food security, impacting agricultural yields and threatening livelihoods. Rising temperatures, altered precipitation patterns, and increased frequency of extreme weather events directly challenge traditional farming methods. The vulnerability of conventional agriculture, heavily reliant on monocultures and synthetic inputs, is starkly apparent. This highlights the urgent need for resilient, climate-smart agricultural approaches. Agroecology, with its emphasis on biodiversity, ecological balance, and integrated systems, offers a powerful set of tools to address these challenges.

Its strength lies not just in its ability to mitigate climate change but also in its capacity to build agricultural resilience against its impacts. Unlike conventional agriculture, which often exacerbates climate issues through greenhouse gas emissions and resource depletion, agroecology actively works towards carbon sequestration, improved water management, and increased biodiversity – all key aspects of climate change mitigation and adaptation. Consider, for example, the critical role of soil health in climate change mitigation. Healthy, diverse soils are significantly better at sequestering carbon than degraded soils found in many conventional farms. Agroecological practices, such as cover cropping, no-till farming, and the incorporation of organic matter, directly enhance soil carbon storage capacity. This action not only helps reduce atmospheric carbon dioxide levels but also improves soil fertility and water retention, making farms more resilient to drought and extreme weather.

Furthermore, agroecology promotes biodiversity both above and below ground. Diverse cropping systems, integrated pest management strategies and the inclusion of beneficial plants all enhance the resilience of farms to various stresses including pests and diseases, often made worse by changing climate conditions. This inherent resilience minimizes the need for external inputs, such as pesticides and fertilizers, which often have high carbon footprints. A diversified farming system is less susceptible to widespread crop failure caused by climate shocks, a critical advantage in an era of unpredictable weather patterns.

Water resource management is another crucial area where agroecology provides solutions. Techniques like water harvesting, efficient irrigation systems, and drought-tolerant crop selection are all integral parts of agroecological approaches. These methods help reduce water consumption and enhance the efficient use of available water resources, especially important in regions facing increasing water scarcity due to climate change.

Beyond the technical aspects, agroecology also addresses the social and economic dimensions of climate change adaptation. By strengthening local food systems and promoting farmer knowledge and innovation, it builds community resilience and reduces the vulnerability of farmers to climate-related shocks. Supporting local food production also reduces the carbon footprint associated with long-distance transportation, further mitigating

climate change impacts.
Implementing agroecological practices requires a shift in mindset and approach. Farmers need access to appropriate training, support, and resources to transition successfully. Policies and incentives are essential to encourage and enable widespread adoption, creating the conditions for large-scale agroecological transformation. Research is also critical to advance our understanding of the many facets of agroecological systems and to develop new, climate-resilient approaches.
Ultimately, transitioning to agroecology is not merely a matter of adopting specific practices but a fundamental change in how we view agriculture. It's a move away from treating agriculture as a purely extractive industry to one that recognizes the crucial links between agricultural production, ecological health, and social well-being. This transition is crucial, not just for responding to the immediate challenges of climate change, but for building a more sustainable and resilient food system for generations to come. The potential of agroecology to create a more secure and equitable future is immense, but realizing this potential requires collective action and commitment. This necessitates a collaborative effort involving farmers, researchers, policymakers, and consumers alike, working together to shape a future where food production and environmental stewardship are inextricably linked. The future of food security is undeniably intertwined with the adoption of climate-smart agricultural systems. Agroecology offers a viable pathway toward achieving this crucial goal, fostering resilience and addressing the complex challenges of climate change.

5.1 Climate Change Impacts

Climate change presents a formidable challenge to global food security, significantly impacting agricultural productivity and threatening the livelihoods of millions. Rising temperatures, altered rainfall patterns, and increased frequency of extreme weather events like droughts, floods, and heatwaves are already disrupting traditional farming practices and causing widespread crop failures. This necessitates a fundamental shift in how we approach food production, making the transition to climate-resilient agroecological systems imperative.

The effects of climate change are multifaceted and deeply interconnected. Increased temperatures, for example, not only directly affect crop yields by impacting plant growth and development but also exacerbate water stress, increasing the demand for irrigation. This increased demand, combined with altered rainfall patterns, puts immense pressure on water resources, potentially leading to water scarcity and conflicts over water rights. Simultaneously, more frequent and intense heatwaves can cause significant damage to crops, livestock, and infrastructure, leading to substantial economic losses for farmers and entire communities.

Beyond the immediate impacts on crop yields, climate change also influences pest and disease dynamics. Warmer temperatures and altered precipitation patterns can favor the spread of pests and diseases, increasing the risk of crop damage and the need for pesticide use, which poses additional environmental and health risks. Furthermore, changes in temperature and precipitation patterns can alter the distribution and abundance of pollinators, which is crucial for the reproduction of many crops, affecting pollination services and impacting crop yields. Therefore, addressing these interconnected challenges requires holistic solutions.

Agroecological approaches offer a powerful toolkit to mitigate and adapt to these challenges. Diversifying crops, for example, can create more resilient systems that are less vulnerable to the impacts of climate change. A diverse range of crops is less likely to be impacted by the same extreme weather events or pests simultaneously, creating a buffer against losses. The practice of crop rotation can improve soil fertility and resilience, fostering better nutrient utilization and making plants more drought-tolerant. Similarly, agroforestry systems which integrate trees into farming systems, can provide shade, reduce soil erosion, and improve water infiltration, creating a microclimate that is more conducive to crop growth, especially during periods of heat stress.

Improving soil health is critical for enhancing resilience to climate change impacts. Healthy soils have a higher water-holding capacity, improving drought tolerance, and are better able to sequester carbon, helping to mitigate climate change. Practices such as cover cropping, no-till farming, and the use of compost and manure help to improve soil structure, fertility, and water retention, making them crucial components of climate-resilient agricultural systems. Furthermore, integrating livestock into farming systems can improve soil fertility through manure and improve nutrient cycling in the system while increasing the overall farm biodiversity. The holistic approach offered by agroecology allows for the combined benefits of enhanced resilience to extreme weather events and the mitigation of climate change through improved soil health.

The implementation of climate-smart agroecology necessitates a shift in both mindset and practices. It requires a deep understanding of local ecosystems, adapting strategies to suit specific environmental conditions and cultural contexts. This participatory approach involves engaging local communities, farmers, and researchers in collaborative efforts to design and implement locally adapted solutions. Knowledge sharing and peer-to-peer learning networks play a crucial role in enabling widespread adoption of these practices. Collaboration amongst stakeholders is vital for promoting adaptive capacity, creating resilience amongst farming communities, and ensuring the equitable distribution of resources.

Addressing the impacts of climate change on food production requires a multi-pronged approach. It is essential to integrate climate change adaptation and mitigation strategies into all aspects of agricultural planning and policy-making. This implies recognizing agroecology not just as a farming system but as a framework for building resilience in our food systems. Investing in research and development to create climate-resilient crop varieties, enhancing early warning systems for extreme weather events, and building the capacity of farmers to adapt to climate change are all crucial steps in ensuring food security in a changing world. The transformation towards a more sustainable and climate-resilient food system is a shared responsibility, requiring collaboration amongst governments, researchers, communities, and farmers. A significant shift in policy, infrastructure, and resource allocation is crucial for creating a world where food production is both environmentally sustainable and socially equitable.

5.2 Mitigation & Adaptation

Climate change presents a significant threat to global food security, impacting crop yields, livestock production, and overall food availability. Agroecological practices, however, offer a powerful suite of both mitigation and adaptation strategies, crucial for building resilience in the face of these challenges. This makes the adoption of such practices essential, not just for the environment, but for global food security in the long run. We need to look to improve the agricultural practices for the benefit of the future generation.

One primary mitigation strategy involves enhancing carbon sequestration in soils. Conventional agriculture often leads to soil degradation and carbon release, contributing to climate change. In contrast, agroecological practices like cover cropping, no-till farming, and the incorporation of diverse organic matter—things like compost and manure— significantly improve soil health. This leads to increased carbon storage, effectively mitigating greenhouse gas emissions and helping fight against climate change. The increased carbon content improves soil structure, nutrient retention, and water infiltration and can have a positive impact on agricultural yield.

Adaptation strategies focus on building resilience to climate change impacts. These impacts include more frequent and intense droughts, floods, and extreme temperatures. Diversifying crops and livestock breeds is a key adaptation measure. Using drought-tolerant varieties for drier regions, or flood-resistant crops in areas prone to inundation, is a practical application of this. Intercropping, where different crops are grown together, can improve resilience by creating diverse canopies for increased water retention and pest suppression, bolstering adaptive capacity to climate variability. It's all about improving the ecosystem services by enhancing the environmental health.

Integrated pest management (IPM) represents another crucial adaptation strategy. As climate change alters pest distributions and behavior, reliance on chemical pesticides can become even more problematic. IPM, which integrates biological control, cultural practices, and minimal pesticide use, offers a more sustainable and resilient approach. IPM reduces reliance on pesticides which are harmful to the environment, thereby minimizing the damage done. Moreover, by promoting biodiversity, IPM helps create more stable and productive agricultural systems that can better withstand environmental stress. It fosters the balance in the ecosystem.

Furthermore, agroforestry systems—integrating trees and shrubs into agricultural landscapes—can play a significant role in both mitigation and adaptation. Trees provide shade, reducing water stress on crops during droughts and helping to alleviate the effect of increased temperatures. They can also serve as windbreaks, protecting crops from damage during storms. This offers several benefits to the environment and food security. The trees themselves sequester carbon, mitigating climate change.

Water management is paramount in adapting to climate change. Agroecological practices such as rainwater harvesting, efficient irrigation techniques (like drip irrigation), and mulching help conserve water resources and make farms more resilient to droughts. These strategies reduce the negative environmental impacts associated with conventional agricultural practices and help in conserving water resources effectively.

Investing in research and development is essential to further enhance agroecological practices for both mitigation and adaptation. This includes breeding climate-resilient crops, improving IPM strategies for novel pests, and developing more effective soil management techniques adapted to changing climatic conditions. The potential of agroecology is huge, yet to be fully tapped.

Community engagement and education are crucial for successful implementation. Building awareness about the benefits of agroecology, supporting farmers in transitioning to sustainable practices, and fostering community-based adaptation strategies are all key steps. Empowering local communities to participate in shaping their agricultural systems ensures the long-term success and sustainability of initiatives and helps in food security.

The transition to agroecology is not just an agricultural shift; it's a fundamental transformation towards more sustainable, climate-resilient food systems. By integrating mitigation and adaptation strategies, we can create more robust agricultural systems that not only feed the world but also help mitigate the impacts of climate change and enhance the health of the environment. The success of this transition depends on collective action, policy support, and a commitment to long-term sustainability. The combination of actions is vital for the long-term benefit.

6 Policy and Governance

Effective policy and governance are paramount for the widespread adoption of agroecological practices. Without supportive frameworks, the transition to more sustainable agricultural systems will remain fragmented and slow. Current agricultural policies often prioritize large-scale, monoculture farming, heavily reliant on synthetic inputs, a stark contrast to agroecology's principles of diversity and ecological balance. This necessitates a significant shift in policy direction.

Incentive programs can significantly encourage farmers to embrace agroecological techniques. Direct financial support, such as subsidies for cover cropping, agroforestry, or integrated pest management, can offset the initial investment costs and potentially lower long-term production expenses. Similarly, tax breaks and grants can attract investment in research and development specific to agroecological innovations. These incentives should be tailored to specific contexts, acknowledging regional differences in climate, soil types, and existing farming practices. A nuanced approach is vital to achieve widespread impact.

Regulations play a crucial role in shaping agricultural practices. Restricting or taxing the use of harmful pesticides and fertilizers, while simultaneously promoting biodiversity and soil health, can level the playing field, incentivizing sustainable methods. Regulations could also prioritize agroecological practices in government procurement and public land management, increasing demand and driving market growth. It's important to design regulations that are both effective and realistic, avoiding overly burdensome requirements that discourage farmers from participating. Collaboration between policymakers, scientists, and farmers themselves is key to formulating policies that are both supportive and achievable. Successful regulation requires careful consideration of both ecological and socioeconomic factors.

Beyond financial incentives and regulations, fostering a collaborative environment between government agencies, research institutions, and farmer organizations is vital for the success of agroecological transition. This collaborative framework can facilitate the sharing of knowledge, best practices, and technological innovations. Regular dialogue and effective communication channels can help ensure that policies reflect the needs and realities of farmers on the ground. This participatory approach builds trust and promotes a sense of ownership amongst all stakeholders, fostering greater commitment to the transition. The collective strength of diverse perspectives results in more effective and equitable policies.

Supporting agroecological research is crucial for its long-term success. Government funding should be directed toward studying the effectiveness of agroecological techniques in diverse settings, including research on biodiversity, soil health, and pest management. This evidence-based approach will strengthen the argument for agroecological adoption and inform policy decisions. Investing in educational programs for farmers and extension agents is equally important, ensuring knowledge transfer and facilitating the practical implementation of agroecological methods. Access to information and hands-on training will be crucial in boosting farmers' confidence and capabilities. The investment in research should be ongoing and adaptable, considering the evolving challenges and opportunities in agriculture.

Furthermore, successful policy must consider the social and economic dimensions of food systems. Policies supporting fair prices for farmers, promoting local food markets, and ensuring access to healthy food for all populations will strengthen the sustainability of agroecological systems. The transition to agroecology needs to support food security and justice, ensuring that vulnerable populations are not negatively impacted. This requires a comprehensive approach that addresses not just the environmental aspects of food production but also its social and economic consequences. Just and equitable policies will result in more stable and sustainable food systems. These considerations should be incorporated into all aspects of policy development and implementation.

Finally, a long-term vision is needed. Policies supporting agroecology shouldn't be short-term fixes, but rather components of a broader strategy for sustainable food systems. This strategy should encompass policies that address climate change, biodiversity loss, and food security. Integrating agroecology into a comprehensive framework ensures its long-term resilience and contributes to a more sustainable future for all. This holistic approach is essential for ensuring lasting impact and building a resilient, equitable, and environmentally sound food system. The journey towards widespread adoption of agroecology requires sustained commitment and a long-term vision.

6.1 Agricultural Policies

Effective agricultural policies are crucial for fostering the widespread adoption of agroecological practices. These policies must move beyond simply supporting conventional farming and actively incentivize the transition to more sustainable systems. This requires a multi-pronged approach that addresses economic, environmental, and social considerations simultaneously. Farmers, often operating on tight margins, need substantial financial support to adopt new methods.

Direct financial assistance, such as grants or subsidies, can help cover the initial costs of transitioning to agroecological practices. These costs might include acquiring new equipment, investing in soil testing, or engaging in training programs. Subsidies should be designed to reward specific agroecological practices, such as cover cropping, crop rotation, and the integration of livestock, rather than simply providing blanket payments.

Beyond direct financial support, policies can create more favorable market conditions for agroecological products. This might involve establishing certification standards that guarantee the quality and sustainability of these products, thereby increasing consumer confidence and demand. Government procurement policies can also play a role, prioritizing the purchase of agroecologically produced food for public institutions such as schools and hospitals. This creates a guaranteed market for farmers and further incentivizes the transition.

Education and training are essential components of a successful policy framework. Farmers need access to information and resources that enable them to successfully implement agroecological techniques. This includes providing practical training on various aspects of agroecology, from soil management and pest control to marketing and business planning. Investing in agricultural extension services and creating farmer-to-farmer learning networks are crucial steps in this regard.

Furthermore, agricultural policies should address the broader social and environmental context of food production. This includes supporting the development of local food systems, which can foster greater food security and community resilience. Policies should encourage the creation of farmers' markets, community-supported agriculture (CSA) programs, and other initiatives that connect farmers directly with consumers. Similarly, policies should incorporate biodiversity conservation measures, ensuring that farming practices contribute to the protection of ecosystems and habitats.

Environmental regulations can play a crucial role in shaping agricultural practices. For example, policies that restrict the use of harmful pesticides and fertilizers can drive the adoption of integrated pest management (IPM) techniques and the use of organic amendments. Regulations that protect water resources, such as limiting irrigation withdrawals or promoting water-efficient technologies, are also crucial. Strong environmental regulations are essential in curbing environmentally damaging practices and fostering environmental stewardship.

The design and implementation of effective agricultural policies require collaboration among various stakeholders, including farmers, researchers, policymakers, and consumers. This participatory approach ensures that policies are relevant, responsive, and inclusive. Regular monitoring and evaluation of policy effectiveness are also critical to adapt policies and ensure they remain effective in supporting the transition to agroecology over the long term. Successful policy implementation relies heavily on clear communication, adequate resources, and strong political will.

Ultimately, successful agricultural policies must balance economic viability with environmental sustainability and social equity. They must support farmers in their transition to agroecology while protecting the environment and enhancing community well-being. A holistic, multi-faceted approach is required, one that considers the entire food system from farm to table. This includes everything from research and development of new agroecological techniques to the creation of supportive market conditions and equitable access to resources. Only through such comprehensive strategies can we hope to unlock the full promise of agroecology. The challenge lies in implementing these strategies effectively, ensuring their widespread adoption, and securing the long-term viability of agroecological farming systems. The future of food security and environmental sustainability depends on it. A failure to act decisively will have profound and far-reaching consequences, impacting not just agricultural practices but the health of our ecosystems and the well-being of generations to come.

6.2 Supporting Agroecology

Governments hold considerable power in shaping agricultural landscapes. Policies directly influence farmer choices, impacting the adoption of agroecological practices. Subsidies favoring conventional farming often discourage the transition, creating a significant barrier for those seeking more sustainable approaches. Reversing this trend requires a fundamental shift in policy priorities, favoring agroecological methods with targeted incentives. This means redirecting funds away from practices detrimental to the environment and investing in programs promoting biodiversity, soil health, and water conservation. Furthermore, policymakers can create regulations that incentivize ecological stewardship by farmers, making environmentally friendly farming more financially viable. This may involve establishing minimum standards for soil health or limiting the use of harmful pesticides, creating a level playing field for agroecological farms while discouraging environmentally damaging practices. Consider the potential for carbon credits based on practices that sequester carbon in soil. This not only addresses climate change but also economically supports farmers adopting these crucial conservation techniques. The complex web of regulations and subsidies must be carefully navigated, necessitating a multi-faceted approach. It's not enough to simply remove harmful subsidies; positive incentives must be created to attract farmers to agroecological methods.

Education plays a crucial role. Farmers need access to comprehensive training and support to successfully transition to agroecology. This involves a multifaceted program encompassing practical on-farm demonstrations, hands-on workshops, and educational resources tailored to different contexts and farming systems. Creating networks connecting experienced agroecological farmers with newcomers fosters peer-to-peer learning, a potent mechanism for knowledge transfer and the development of locally-relevant techniques. Furthermore, educational initiatives should reach beyond farmers to include consumers, empowering them to make informed choices about food systems and demand agroecological products. The dissemination of information needs to happen through diverse channels, engaging stakeholders at all levels—from schoolchildren to policymakers. This broad-based approach ensures that support for agroecology is not just a technical matter but becomes embedded within a broader cultural shift. It should also involve fostering collaboration between research institutions, educational bodies, and farming communities, generating a continuous flow of innovations and best practices. A combination of formal training, participatory learning processes, and ongoing support can maximize the impact of educational initiatives, effectively promoting the wider adoption of agroecological practices.

Access to finance is a major hurdle for many aspiring agroecological farmers. Conventional lending institutions often prioritize large-scale, intensive farming operations, overlooking the smaller-scale, more diverse systems characteristic of agroecology. Creating alternative financing mechanisms is essential. This could involve developing microfinance schemes tailored to the specific needs of agroecological farmers, or establishing funds specifically dedicated to supporting the transition to sustainable agriculture. Furthermore, fostering partnerships between financial institutions and agroecological organizations can bridge the gap between capital needs and sustainable agricultural practices. These collaborations can help design loan products that reflect the unique characteristics of agroecological farms, such as longer-term investments and risk mitigation strategies. Considering innovative financing tools, such as crowdfunding or community-based investment models, can provide additional avenues for funding. A diversified approach to finance should account for farmers' specific needs, risks, and capacities, effectively enhancing accessibility to capital and fostering greater adoption of agroecological methods. Such an approach is vital for ensuring financial viability and long-term sustainability.

Research and development are vital in constantly improving and expanding agroecological techniques. Ongoing research is needed to explore and refine existing practices, tailoring them to diverse agro-ecological zones and farming contexts. This includes developing new crop varieties that are resilient to climate change and pests, and enhancing our understanding of soil biology and its role in nutrient cycling. Furthermore, robust research networks are crucial, enabling the sharing of knowledge and best practices across geographical locations. Collaborations between scientists, farmers, and policymakers ensure that research is relevant to real-world challenges and effectively translated into practical applications on farms. A strong emphasis on participatory research, involving farmers as active participants throughout the research process, ensures that solutions are locally relevant and effectively integrated into existing farming systems. This approach fosters a sense of ownership and encourages the uptake of new technologies and practices within farming communities. Investing in long-term research ensures a continuous flow of innovative techniques and knowledge to address the ever-evolving challenges faced by agroecological farmers.

Consumer demand significantly influences farming practices. Increased consumer awareness of the environmental and social benefits of agroecology can drive market demand for agroecological products. This requires targeted communication strategies to raise public awareness of the positive impacts of agroecology on food systems, health, and environmental sustainability. Promoting agroecological products through farmers' markets, community-supported agriculture programs, and direct-to-consumer channels facilitates a direct connection between producers and consumers. Educating consumers about the features and benefits of agroecological products helps to build consumer confidence and increase demand. Such initiatives empower consumers to play an active role in supporting sustainable agriculture, creating a virtuous cycle of demand and supply. Developing clear labeling standards for agroecological products ensures transparency and strengthens consumer trust. By supporting agroecology through conscious consumer choices, consumers can actively contribute to building a more sustainable and resilient food system.

6.3 Incentives & Regulations

Effective policy design is crucial for the widespread adoption of agroecological practices. Incentives, particularly those focusing on long-term ecological benefits rather than short-term yields, can drive a significant shift. Imagine a system where farmers receive direct payments for practices that improve soil health, biodiversity, and carbon sequestration—payments that reward them for building ecological resilience into their farms, not just for maximizing immediate harvests. This type of policy shift necessitates a clear understanding of the ecological services provided by agroecological farms, allowing for precise and effective payments based on verifiable outcomes.

Regulations, conversely, play a vital role in establishing minimum standards and preventing practices that undermine agroecological goals. Consider the potential impact of regulations limiting the use of synthetic pesticides and fertilizers, forcing a transition to more sustainable and environmentally friendly alternatives. While some might argue that such regulations stifle innovation or increase costs, a well-designed regulatory framework should prioritize both environmental protection and the long-term viability of farming systems. Balancing these competing interests requires thoughtful consideration and careful consultation with all stakeholders, ensuring a fair and equitable transition. This includes actively engaging with farmers to address concerns and foster collaboration.

For instance, a government could offer tax breaks or subsidies for farmers who adopt specific agroecological practices, such as cover cropping or no-till farming. This financial assistance can offset the initial investment costs and encourage wider adoption. Furthermore, regulations that prohibit certain harmful farming practices, while providing support for transitioning to agroecological alternatives, ensures a gradual and manageable shift to more sustainable approaches. This is particularly relevant when dealing with intensive farming systems and the use of high volumes of external inputs. It avoids imposing insurmountable changes, allowing farmers to adapt and evolve without causing significant economic disruption.

The creation of farmer cooperatives and supporting their access to fair markets also plays a key role in promoting agroecological practices. These cooperatives help farmers collectively negotiate better prices for their products and access resources that might otherwise be unavailable to them individually. This strengthens their position in the market and provides an economic incentive to adopt sustainable farming methods. Furthermore, establishing clear labeling standards that distinguish agroecological products from conventional ones can significantly boost consumer demand and increase the market value of these sustainably produced goods. This transparency and consumer awareness empowers consumers to actively support sustainable practices.

Education and outreach initiatives are also vital for promoting agroecology. These initiatives would inform both farmers and consumers about the benefits of agroecology, empowering them to make informed choices. Workshops, training programs, and educational materials can provide farmers with the knowledge and skills necessary to successfully implement agroecological practices. Simultaneously, educating consumers about the environmental and social benefits of agroecological products helps foster demand for these sustainable products, creating a more vibrant market for agroecological farms. This dual-pronged approach, focusing both on supply and demand, establishes a foundation for long-term success.

Building a supportive policy environment for agroecology requires a holistic approach. It involves a combination of incentives that reward farmers for adopting sustainable practices and regulations that prevent harmful activities. Moreover, ensuring market access for agroecological products and providing adequate support through education and outreach initiatives is essential. This comprehensive strategy fosters the transition towards more sustainable food systems, ensuring a food supply that is both environmentally friendly and economically viable for the long term, supporting rural communities and preserving the natural resources they depend on. The success of agroecology ultimately hinges on policy that balances ecological concerns with economic realities, fostering a pathway toward a more sustainable future for food production.

7 Future of Agroecology

The future of agroecology hinges on a multifaceted approach encompassing research, scalability, and global food system transformation. We need robust research programs focusing on climate-resilient crops, improved pest and disease management strategies tailored to diverse agro-ecological zones, and innovative technologies that enhance efficiency while minimizing environmental impact. This research must be participatory, involving farmers and local communities in the design and implementation of studies, ensuring the solutions developed are both effective and culturally appropriate. This collaborative approach ensures that the knowledge generated is readily adopted and relevant to the needs of those who will implement it.

Scaling up agroecological practices requires a strategic blend of policy interventions, financial incentives, and capacity building initiatives. Governments need to implement policies that actively incentivize farmers to adopt agroecological methods, providing subsidies, tax breaks, or other forms of financial assistance. This support might include direct payments for ecosystem services, such as carbon sequestration or biodiversity enhancement, rewarding farmers for their contribution to environmental sustainability. Equally crucial is investing in farmer training programs that equip individuals with the skills and knowledge needed for successful agroecological transition. Beyond individual farmers, supporting collaborative networks and farmer cooperatives will enable a faster, wider scale adoption of agroecological systems.

Transforming global food systems requires a shift in mindset and actions beyond simply adopting farming practices. This implies a conscious movement towards localized food production systems, reducing reliance on long, unsustainable supply chains. Strengthening local and regional food markets through initiatives such as farmers' markets, community-supported agriculture (CSA) schemes, and community food hubs can build resilience within food systems. These initiatives foster local economies, and support food sovereignty, empowering communities to control their food production and distribution. Furthermore, addressing issues of food waste and loss is crucial; significant amounts of food are wasted throughout the entire supply chain, from farm to plate. By tackling waste and implementing efficient storage, distribution, and processing methods, food security can be improved substantially.

Ultimately, the future of agroecology depends on integrating ecological, economic, and social considerations. A truly sustainable food system requires equitable access to resources and opportunities for all stakeholders, including smallholder farmers, who form the backbone of many food systems. This necessitates addressing issues of land tenure, access to credit, and market access, ensuring that all benefit from the transition to sustainable agriculture. It also demands fostering collaborations between researchers, policymakers, farmers, and consumers, creating a cohesive framework supporting the widespread adoption of agroecological principles. Only through this integrated and inclusive approach can we ensure the long-term success and sustainability of agroecological systems, realizing their full potential to provide food security, environmental stewardship, and community well-being for generations to come. The challenges are significant, yet the rewards – a more sustainable, resilient, and equitable food future – are undeniable. This comprehensive, multi-pronged strategy lays the foundation for achieving a future where food production not only nourishes humanity but thrives in harmony with nature. This integrated vision offers a pragmatic pathway for building a food system capable of meeting the demands of a growing global population while mitigating climate change and conserving natural resources, leaving a legacy for future generations. The future of agroecology isn't merely about technological advancements or policy changes; it's about a fundamental shift in our relationship with food and the land that sustains us.

7.1 Research & Development

The future of agroecology hinges on robust research and development. We need to move beyond anecdotal evidence and embrace rigorous scientific investigation to refine existing practices and develop innovative solutions. This necessitates a multidisciplinary approach, bringing together agronomists, ecologists, soil scientists, social scientists, and engineers. Collaborative efforts are vital; universities, research institutions, and farming communities must work together to address specific challenges within diverse agroecological systems. Funding for such research is paramount, requiring significant investment from governments, private sector entities, and philanthropic organizations. This isn't simply about throwing money at the problem; it's about strategically allocating resources to address the most pressing needs and ensure the long-term viability of agroecology.

Developing climate-resilient crops is critical. Traditional varieties often lack the adaptability needed to withstand increasingly erratic weather patterns. Research should focus on identifying and breeding crops with improved drought tolerance, heat resistance, and flood resistance. This involves studying genetic diversity within existing crop species and exploring techniques like marker-assisted selection and gene editing to accelerate the breeding process. Simultaneously, we need to understand how different agroforestry systems can buffer crops against climate change impacts and enhance resilience.

Precision agriculture offers exciting possibilities. By integrating sensor technology, data analytics, and targeted interventions, farmers can optimize resource use, reduce waste, and enhance productivity. This includes developing tools to monitor soil health, track pest outbreaks, and optimize irrigation strategies. This precision approach can help minimize environmental impacts while increasing the economic viability of agroecological farms. However, accessibility remains a crucial factor; making this technology affordable and user-friendly for smallholder farmers in developing countries is vital for widespread adoption.

Improving pest and disease management within agroecosystems requires a comprehensive understanding of ecological interactions. This necessitates studying predator-prey dynamics, plant-microbe relationships, and the impact of diverse cropping systems on pest populations. Developing effective and environmentally friendly pest control strategies goes beyond simply identifying effective biological control agents; it demands an integrated approach that leverages diverse mechanisms for pest suppression, from habitat manipulation to resistant crop varieties. The focus should be on long-term strategies that minimize reliance on synthetic pesticides and enhance the inherent resilience of agroecological systems.

Effective soil management is the cornerstone of agroecology. Research should focus on enhancing soil fertility using organic amendments, cover cropping, and no-till farming practices. We must deepen our understanding of soil carbon sequestration and its role in mitigating climate change. This involves quantifying the carbon storage potential of different agroecological systems and developing strategies to maximize carbon sequestration while also boosting soil health. In addition, we must study the long-term impacts of various soil management practices on soil structure, nutrient cycling, and water retention capacity.

Further research is needed to assess the socio-economic impacts of agroecology. This involves studying the livelihoods of farmers, the economic viability of agroecological farms, and the nutritional benefits of agroecological produce. Analyzing market access and consumer preferences are crucial for guiding the development of sustainable and equitable food systems. This research must go beyond simple economic assessments; it should investigate the broader societal benefits of agroecology, including its contribution to community resilience, food security, and social equity. The long-term sustainability of agroecology depends on demonstrating its economic and social viability in diverse contexts.

Finally, we need to develop robust monitoring and evaluation frameworks. This involves establishing standardized methodologies for assessing the ecological, economic, and social performance of agroecological systems. This requires building a network of researchers and practitioners who can collect and analyze data, identify best practices, and disseminate knowledge widely. Long-term monitoring is crucial for evaluating the long-term success of agroecological initiatives and for making data-driven decisions about future research priorities. The aim is to build a robust evidence base to support the widespread adoption and implementation of agroecological principles globally.

7.2 Scaling Up Agroecology

Scaling up agroecology requires a multi-pronged approach, moving beyond isolated projects to widespread adoption. This isn't simply about increasing the number of agroecological farms; it demands systemic change impacting policy, education, and market structures. We need to shift from viewing agroecology as a niche practice to recognizing it as a vital solution for global food security and environmental sustainability.

Successful scaling necessitates a comprehensive strategy addressing crucial barriers. Financial accessibility is paramount; many farmers lack the initial capital for transitioning to agroecological methods. Innovative financing mechanisms, like micro-loans tailored to agroecological practices or government subsidies prioritizing sustainable farming techniques, are crucial. This investment should include support for farmer training and mentorship programs. These programs should not just focus on technical skills, but also foster a community of practice, allowing farmers to share knowledge and experiences.

Furthermore, market access remains a significant hurdle. Consumer demand for agroecologically-produced food needs to increase, and effective marketing strategies are essential to bridge the gap between producers and consumers. This means building strong local food systems, creating clear labeling standards for agroecological products, and fostering relationships between farmers and retailers committed to sustainability. Initiatives such as farmers' markets, community-supported agriculture (CSA) programs, and direct-to-consumer sales channels can play a crucial role.

Policy and governance play a vital role in facilitating scaling. Governments must actively support agroecology through various policy instruments. These could range from creating supportive regulations, such as reduced taxes for sustainable agricultural inputs, to incentivizing agroecological practices with direct payments or grants. Investment in agroecological research and development is equally important, as it helps to refine existing techniques and develop new ones. This includes funding research into climate-resilient crops and integrated pest management strategies specifically suited to various agro-ecological zones.

Education and outreach are critical components in the scaling process. Integrating agroecological principles into agricultural curricula at both secondary and tertiary levels is essential to cultivate a future generation of farmers, researchers, and policymakers dedicated to sustainable agriculture. Public awareness campaigns are also essential to educate consumers about the benefits of agroecology and to encourage their support for sustainable food systems. These campaigns need to be engaging and accessible to diverse audiences, utilizing various channels of communication to maximize their impact.

Technological advancements can also contribute significantly to scaling agroecology. Precision agriculture technologies, like sensors and drones for monitoring crop health and optimizing resource use, can improve the efficiency and productivity of agroecological farms. However, it's crucial to ensure that these technologies are accessible to smallholder farmers and designed to complement rather than replace traditional agroecological knowledge. We need to find a synergy between technological innovation and the time-tested wisdom of generations of farmers.

Beyond individual farms, scaling requires a holistic approach encompassing entire landscapes and food systems. This includes promoting agroecological practices in specific geographical regions by fostering collaborations between farmers, researchers, and local communities. The creation of agroecological zones, where several interconnected farms employ common principles, can generate significant benefits. By cooperating on resource management, sharing infrastructure, and developing common market channels, these farms can enhance their overall sustainability and resilience.

Finally, robust monitoring and evaluation systems are crucial to track progress, identify challenges, and adapt strategies as needed. Regular assessment of the environmental, social, and economic impacts of agroecological initiatives allows for improvements and better decision-making. Transparency and data sharing are key to ensuring accountability and learning from both successes and failures. The data generated can also help in advocating for policy changes and securing further funding for agroecological projects. Scaling up agroecology is not merely a matter of expanding existing models; it's about creating transformative change in our food systems, and this transformation demands a combined effort from all stakeholders.

7.3 Global Food Systems

Understanding global food systems requires recognizing their immense complexity. They are not simply about producing food; they encompass intricate webs of production, distribution, consumption, and waste management, all interwoven with social, economic, and environmental factors. These systems, currently largely reliant on industrial agriculture, face mounting pressure from population growth, climate change, and resource depletion. This creates a critical need for transformative change.

Shifting towards agroecological practices offers a powerful pathway for addressing these challenges. Agroecology moves beyond simply maximizing yields; it prioritizes the long-term health of ecosystems and the well-being of communities. Instead of viewing farms as isolated units, agroecology emphasizes their integration within broader landscapes. This integrated approach recognizes the interconnectedness of various factors, such as biodiversity, soil health, water availability, and human communities.

A core component of transforming global food systems lies in fostering resilient, localized food production. This involves supporting farmers who embrace agroecological principles and enabling them to thrive. It means providing access to resources, training, and markets that prioritize sustainability and fairness. For example, encouraging the growth of farmers' markets and community-supported agriculture initiatives directly connects consumers with local producers and improves food security within communities.

Simultaneously, a shift towards more diverse diets is crucial. Industrial food systems have promoted monocultures, leading to nutrient deficiencies and increased vulnerability to pests and diseases. Diversifying crop production and promoting consumption of a broader range of foods enhances dietary quality and nutritional security. This also strengthens the resilience of food systems by reducing their reliance on a limited number of crops.

We must also grapple with the environmental footprint of our food systems. The industrial model's heavy reliance on synthetic fertilizers and pesticides poses significant risks to human health and environmental quality. Agroecological methods, conversely, prioritize ecological integrity and minimize environmental harm through techniques like integrated pest management and reduced reliance on chemical inputs. The resulting improvement in soil health, water conservation, and biodiversity plays a vital role in mitigating climate change.

Global food systems are fundamentally political. Policies at local, national, and international levels heavily influence production practices, market access, and consumption patterns. Supportive policy frameworks that incentivize agroecological transitions are essential. This requires initiatives that reward environmentally friendly practices, encourage farmers' adoption of sustainable methods, and address issues like land access and equitable market participation. Furthermore, international cooperation is critical for harmonizing food standards and supporting developing nations in their transition to sustainable agricultural practices.

Moving towards a genuinely sustainable global food system requires a significant cultural shift. This means reevaluating prevailing societal values concerning food production, consumption, and waste. Education and awareness campaigns play a crucial role in promoting mindful consumption and reducing food waste. These changes also necessitate re-evaluating the societal valuation of food; fairly compensating farmers, and supporting agricultural practices that place long-term ecological and social well-being above short-term profit maximization.

Ultimately, the transition to sustainable global food systems is not just an agricultural challenge, it is a fundamental issue of social justice and environmental sustainability. Agroecology offers a holistic approach, integrating ecological principles with social equity and economic viability. Embracing this transformative shift promises a future where food production nourishes both people and the planet, ensuring food security and environmental well-being for generations to come. The path forward requires a multifaceted approach incorporating technological innovation, supportive policies, transformative education, and a fundamental shift in our collective values regarding food and its production. This global endeavor demands collaboration amongst researchers, policymakers, farmers, and consumers. Only through collective effort can we create a resilient and sustainable food system for a healthier future.

www.ingramcontent.com/pod-product-compliance
Lightning Source LLC
LaVergne TN
LVHW022022200926
848481LV00007B/2484
9798348548087